EARTH and TRE
in WAR and PEACE

60 Golden Years - Volume 2

Maurice H. Sanders

CORTNEY PUBLICATIONS

This book is dedicated to the late John Evans, man of steel, lavish in his friendship with those of his ilk.

ISBN 0 904378 51 9

Published by
Cortney Publications,
57 Ashwell Street, Ashwell,
Baldock, Herts SG7 5QT
(Tel: 01462 742185)

Word Processing J. Haerle, Ashwell, Herts.
Page Layout by *Olive*Press Ltd, Stotfold, Hitchin, Herts.
Printing and Binding by Henry Ling Ltd, Dorchester, Dorset.

Contents

Acknowledgements

Heartfelt thanks go to the contributors of this book who have patiently helped make it possible: to Dave Gillow my photographic advisor for all my books; Geoff Tegerdine for text preparation and Peter Allen who collated the captions and much, much more; to all my devoted friends, including Ben Hinton who has introduced me to so many great characters in timber over twenty years. Finally, my thanks to my publisher, Norman Gurney, who in 1983 bravely took on the unknown qualities of timber haulage and an uneducated timber carter, bearing only action snapshots and experience of a way of life hitherto not previously seen in print in this unique way.

My royalties for Volume 2 are destined, as for Volume 1, to the Luton Branch of The Samaritans, who, with other branches, try to ease the depth of human misery in rural Britain, especially in farming communities. MAURICE H. SANDERS

The Round Timber Club

After seven years, for the want of commitment and dedication among some of its members the RTC has sadly been disbanded, or as I prefer to think, put on the "Back Burner" because that dedicated little band of enthusiastic zealots have an inbred nature of "Timber Person" resilience. But, you cannot disband the new and deep cherished friendships so many of us have forged. Let us heartily thank and rejoice in the "Trojan" efforts of the faithful few who gave and gave, time after time. Phil Wallace's "Mats and Millys" gatherings prosper and I know other friends meet up to "air their kit" informally. The Phoenix of the RTC may well rise again in some new form. I understand it was the unanimous and most generous wishes of the members, that the surplus funds (an estimated £1,500) be donated to my charity - The Luton Branch of the Samaritans and I thank you all on their behalf.

If you wish to go on the mailing list should a new venture come to pass my publisher whose address is within this book would forward on to me. To have shared a rich fellowship with so many of you has been a great joy, privilege and pleasure. MAURICE H. SANDERS

Glossary

A.P.F. - Association of Professional Foresters
Artic - Articulated lorry
Cats - D2, D4, D6, D7, and D8 various models of Caterpillar tractors.
Cube - 25 cubic feet of hardwood weighing about a ton
Double Heading - the practice of coupling a second tractor unit for extra power on steep hills
R.B. - Ruston Bucyrus make of excavator
Skidder - Large 4 x 4 tractor winch
Tushing - local name for winching timber

A Tribute to the late John Evans

John Evans was the only reader among thousands to make time to write kind words about all seven of my books and two videos produced for charity, one of which featured in his yard with his old vehicles and drivers which took days to arrange and film. His 'Bit-of-a-Do' gatherings, when friends drawn from a vast cross-section of his business and leisure life met up around his various machines (where many would be running), for a day of nostalgia with lashings of hospitality served by his daughter Carolyn were as unique as John was.

Despite pressure of work he'd find time to visit me, particularly after my wife Helen died. Few would see him as a bereavement counsellor but he was a gifted one, perhaps because he was further down the road that I am now treading. On the last occasion, just prior to his untimely death, he turned up with a load of logs for Christmas: that's the sort of man he was, ever doing good, often by stealth.

John had an infectious smile, a sunny disposition and a great ability to lift one's flagging spirits. I don't think he was a regular church-goer, but he was a shining example of Christian principles to those who are. At the end of his last letter to me he writes, "Your books are still at the top of the pile; when's the next one coming?".

Those privileged to have know him will miss him dearly. Those who were not are sadly the poorer. Edwin Foden's obituary identifies exactly with John's life and their philosophy were much the same. They were great lovers of brass music and each had the hymn 'Abide with me' sung at their funerals. If we do not just remember John Evans, but follow the example of his life and actions, he will not have died in vain.

John Evans with BYN 561 in May 1999 one of only 16 chain drive Fodens built. John's dream was to buy back and restore this tractor previously owned by his father. How very regrettable he was to enjoy so little of his achievements.

On one of John's last visits, he was enthusing over these two photographs. The family has given me permission to use them, although their origin is not known.

Introduction

Operation Crossway Green

For about two years Ben Hinton has said, "Your fan club here (around Worcestershire) keep asking if you're coming up?" A serious request it transpired. On Friday May 8th 1998 this 'Best Friend' as BT 'Friends and Family' system describes him, drove 100 miles to collect me and, with stacks of audio tape and Valium, we set off for his home near Hartlebury, Kidderminster. Here his wife, Marge, awaited me with thoughtful hospitality and, noting Ben's vast itinerary for the next three days, thankfully took over my three-hourly medication programme. After refreshment, within two hours we were heading out to Martly to meet Ray Jackson, a great Caterpillar enthusiast (see Volume I). I required no sleep inducement this night, and over breakfast Ben unfolded the day's plans; starting with a trip over to Droitwich to meet the Gartlands (see Volume I). After a hasty meal, we went on to Colney Green to see a nice International BTD6 re-engined with a BMC diesel much to my surprise, and an ex-WD Ford model WOT6 and the only pole trailer I've ever seen that had a pole adjustment on the front bolster as well as the rear. Then on to Cleobury Mortimer to see Martin Southern and his late father's superb photographs. This was followed by a call in Highley and Button Oak. I'd actually fallen asleep during my last interview and, suspecting 'burn out' of both my tape recorder and me, Ben said, "Better call it a day, we've a heavy programme tomorrow!!" Marge greeted me with overdue pills, culinary delights and so to bed.

The next morning, raring to go – Ben more than me – we set off across the Shropshire Hills, or 'banks' as they call them, our destination Leintwardine, a village near Bucknell, with Ben pointing out the places he'd engaged bottom gear, and various hilltop 'boiling points' of old as we went along, arriving early at Trippleton Farm, home of our host Jonathan Garman. It was rather overwhelming to see almost a convoy of 4x4 vehicles arrive that had been truly 'off the road' – a phrase that means only on the path to their urban cousins. My adrenaline demands rose as I forgot each person's name once introduced to the next. Little did they, or I, know the strategy of the conveyor belt type of interview planned. Guests loaded with photographic proof of their unbelievable, at least to folk outside, lives of timber assembled in one room where , according to the laughter I should have had another tape recorder; but I doubt much of it would have been publishable. Each family in turn came to another room with their own stories to add to the 4¾ hours of tape through which I now had to constantly search to bring a glimpse of their fascinating forest lives. Bernard Owen (who features a bit later) and his wife provided a 5-star running buffet. During these three days I mingled with folk 'I couldn't hold a candle to' as they say. I was and am very proud to be counted as a friend of theirs.

Ben Hinton with 800 cube on S Darke and Sons of Worcester's Foden Mk7 two stroke with Thorneycroft Big Ben tandem back end.
No other man has done more to bring you the characters and photographs in all my books than Ben Hinton.

Bill Southern's Photographs

I covered the Tom Pain story in "Stories of Round Timber Haulage" but these fresh photographs, loaned by his son Martin, belonged to driver Bill Southern seen here with a tidy loaded ERF and young Tom completes the pose looking on in 1957.

Cleobury Mortimer's High Street after a heavy snow fall in 1947. Tom's Cat D2 dozes to make the baker's delivery possible.

This Canadian Chev got through a snow drift so deep workers hung their coats
on top of telegraph poles!

Tom's well turned out Matador in action. Tom Pain faced major surgery in mid-life
yet went on to live to be 95.

4

CHAPTER ONE

Jonathan Garman

Jonathan Garman of Trippleton Farm, Leintwadine, Shropshire is an all-round Plant Contractor who shifts trees, muck and mud from lakes, by twin Fowler Steam ploughing engines and a seven yard scoop bucket if you like and was our host for the day. He showed us round his yard as the other assembled from miles around. The first machine was a modified logging arch, now riding on a pair of scraper front wheels conveniently 9ft legal width, generally drawn by his Cat D4D. A Cat D6 and a 7, and his 22RB dragline were referred to. Bernard Owen's previous ex-B & J Davies 6x6 Matador now cut down to a 4x4 re-engined with a Gardner 5LW. as were others I was to hear of that day. Nearby was a Cat D7E, a rugged 22 ton bread-winning brute. A Barford dump truck on combine wheels, yet another innovation. A Preistman Wolf dragline now serves as a crane. Also an AEC Militant BNH884J now fitted with an Atlas loader by Jon, as some call him, had come straight from a circus, delivered during a winter's night by three men who refused to stay on for refreshment since they had to return to feed the tigers – five gallon drums of KittiKat I presume. A Scania and Ford tractor unit with York tri-axle, adapted again by Jon, transports all the gear, including the Fowlers which were built in Leeds next door to rivals McLarens. The story goes that both of these giants of steam made their engines from scrap thrown over the wall by the other! Jon Garman is the son of the local doctor who had private schooling. His boyish delight was to go on the rounds with his father when in Bucknell. Here the lad enjoyed the joint influence of timber men in the yard of B & J Davies (see page 53 of "Stories of Round Timber Haulage" and page 40 of "Men, Mud and Machines") or equally round the corner in the yard of Dick Woolley, a man who'd apprenticed himself at the Sentinel Works at Shrewsbury. While still a young man, Dick set up threshing kits of his own, an enterprise that later introduced steam to road breaking, making and tar spraying in a most unique manner. At one time Dick owned 32 Sentinel Steamers and encouraged Jon, who bought a well-worn steam roller for £45 when he was 19 years old, with only £60 to his name. His first rolling job came in 1960 at 17/6d per hour. The client found the coal – the rest is history. The mechanical and human occupants of these two yards account for

5

much of the success of this man, who for a whole day put his kitchen, office and rooms at the disposal of these ardent followers. Fellows of the forest who constantly claimed "I've got all your books", little knowing they were featuring in the next at that moment.

Oakley Park 1949, a Woolley Sentinel DG Timber Tractor Reg. No VN 4294 with cleats or spuds for additional traction - a rare sight to behold.

Reg No JB 1655, 1933 "Old Bill" with snow plough at Clun, Shropshire in the 1950s. Driver Jack Walker, mate Bing Bevan.

Jon's Fowler BB1 cable engine No.15333 dredging at Brampton Bryan Park, Herefordshire, 1995

The 7yd dredging scoop.

"The Lincolnshire Imp" was the name of a Ruston steam engine previously low loaded on this ERF new to Lawleys of Churchstoke, Powys.

The Ford "Cargo" well and truly loaded.

At the point of balance.

Some of the fleet lined up in 1991.

Maurice Cox aged 70 has operated this 22RB dragline for years, and could virtually make it sit up and beg.

Running repairs on site. Jon's Vickers 180 at Gore Quarry, Kington, supported by a half cabbed Douglas.

Jon drove this Cat D6B when new.

Jon up to his track tops at Ledbury. It is said that the bricks for the viaduct were all made on site.

Jon's £45 steam roller that started it all when he was 19.

We line up as we leave, backed by Les Morgan's gorgeous Foden Reg No GN 622. Left to right
Joseph Lawley, self, Ben Hinton, Jon Garman, Brian Lawley, Bernard Owens, Mark and Megan
Lawley. Bernard's trusty old Gardner engined Matador reclines at the end. Many really great
timber folk gathered here this day.

CHAPTER TWO

S R & S Davies

While S R & S Davies of Bridgnorth as a family have no connection with their B & J namesakes of Bucknell, they have much in common when it comes to tenacity in timber felling and tushing of big hardwoods. There are three active generations here and I start with Stan who at 86 years of age frowns on ideas of retirement, in spite of a lot of ironmongery in one leg due to slipping down a snowy bank only three years ago in his regular duties of ' measuring up'. Sixty-eight years ago, aged eighteen, he'd teamed up felling with a Cyril Perks of Ludlow. At twenty-four Tom Pain (of Cleobury Mortimer – see page 49 of "Stories of Round Timber Haulage") asked Stan to go to Scotland to manage a contract. Then came the big-un, a 200 acre woodland of massive spruces in Wales. Tushing out these heavyweights with a new Cat D4 and Hyster winch, a name supposed to have come from forest shouts of 'hoist –er' with logging arches in the USA. Stan recalled a brilliant exercise of extraction ingenuity here, where a 500 yard railway line was laid down a steepish gradient and loaded bolster wagons were hauled up by a portable steam engine and winch, and lowered mostly by gravity! Cleverly the timber law of 'ways and means' resulted in a wooden railway type of signal enabling the winch man to know when to pull.

One day Stan was tushing a sixty footer in when loud verbal panic broke out; a cow had wandered onto the line as a wagon descended – a minor and regular hazard to this unsophisticated but successful system. Continuing along the 'Health and Safety' theme, it seems a man broke his finger and his paramedically minded mate cut and split a bit of nut stem, bound the split with insulation tape, then they carried on. On another occasion ' out in the wilds' on hearing of a damaged hand this man said, "I hope you've only broken your finger, I can treat that!" The spirit of our trade is seen in a job with an end-of the-year deadline. Bad weather meant a great loss of timber. Up to a dozen vehicles were being dragged in for loading almost round the clock, with lights in trees powered by a generator and a Foden winched them out as it snowed, working these 'unsocial hours' they'd call in today. Son Ray has some felling photographs that say it all.

One job was in the Savernake forest, with trees at 440 cube. From cross cuts, long bladed two-man 'Danarms' to a five foot bar 'Stihl'. A Fordson early TVO Major with Cooks winch was featured pulling hefty power line cables up a hill midst an array of other modern tackle. The usual breed of extraction man in these parts was called 'Duke' and operated a BTD6 crawler. He slept in Venable's boiler house and in preparation for the Work's Outing put his suit to soak in a stream, held down with stones, washing and drying it. He'd iron it with a sledge hammer!

Today Ray's son Stuart heads the team with a ferocious Canadian built GM engined 'Timber Jack' that will tush out 300 cube at a time. Here is a family business engaged in a most dangerous profession, felling and extracting trees. That's not unusual, but to have three generations active certainly is!

This ERF regularly brought fish down from Midlothian, Scotland returning with Davies' felled timber. Stan is on the left - taken in 1938.

Stan gives the "thumbs up" from a Foden

The years have rolled by, but Stan still measures the trees "big uns" and all.

Young Stuart with their "Timber Jack Skidder". No tree daunts this machine or operator.

The Davies family often worked with the legendary Dick, son of George Barker of Stafford. They say he once put 4,040 cube of green beech (some 60 footers) on 4 loads on this Foden Mk 7 two stroke with 12 speed gearbox. Built to spec. with longer stronger chassis etc.

17

Same load seen from the back end - a hell of a sight this! When Dick Barker overtook you on the road you'd think the long trees would never cease to pass you.

Dick climbed out of the back window of this Matador in 1969.

Marmaduke (Duke) Bowes wisely checks the water on this ERF 5LW with superlow gearbox. A massive load of Oak near Hereford approx 1935.

Foden FG with Cotswold Oak bound for Venables of Stafford in 1965. 65 footers here, hence the amount of butt forward of the front bolster.

Not so much the load, but stuck without crane or winch Dick attached and anchored a chain over the load seen on the front pin, drew forward releasing some logs, than crowbarred the rest off single handed; it's called INGENUITY

Oak for Venables from Leominster in 1967; draw your own conclusions on the cubage!

Brown oak Weobley Hereford Foden model FG 6LW Gardner engine AEC Matador 6 wheel drive, with jib loading before the six and half ton Allis held her down. Taken in 1965.

How to stop a six wheel matador rearing up when loading a big tree.

E.R.F. 5.L.W. Gardner Engine. Loading with three legs at Weston Park in 1965.

A poor photograph, but a must to see Dick loading with his homemade detachable loading jib coupled to his "Allis". He says they've seen a few hairy moments in their day (note the lump of skid like wood that props the jib in action).

Dick with his Allis Chambers HD7 Crawler Tractor arranging a 500 cube Oak ready to load on Apley Estate Bridgenorth. This is the small end of the tree!

George D Barker himself on standard Fordson, petrol engine, Cooke's two speed winch. Purchased new in 1934. Taken at Chepstow in 1961.

1968, **1,000 cube 75 feet** long Spruce for Ludlow, thanks to home built 50 foot pole and ex tank transporter backend. The Barker family were

CHAPTER THREE

John Workman OBE

Described as a doyen of the British forestry scene, John Workman OBE of Gloucestershire never opted for the family sawmill, but took a forester's course at Oxford. At 75 in 1998 he celebrated 50 years' service with the National Trust. For 25 years John travelled all the Trust's woodlands as a Forestry Advisor for England, Wales and Northern Ireland. Working out management plans and programmes, selecting mature trees and thinnings for harvesting planted by his forebears and replanting for posterity. A founder member of the Tree Council and deeply involved with the English Nature Committee, Exmoor National Park and the Cotswolds Area of Outstanding Natural Beauty, the presentation of the unique National Trust's Founder's Medal was richly deserved. This is combined with managing his late father's woodlands, some of which have been made over to the National Trust. Few of us realise our dependence on wood in everyday life. This man's enthusiastic application of

John Workman examines a lofty tree in one of the Workman Woodlands.

stewardship of both scenic and productive sides of woodlands has been tremendous. He has nurtured trees and brought us re-afforestation at its best. Constantly he has seen and applied timber as a great sustainable resource, a lesson we all must learn.

John Workman and his forester David Harris describe this incident on their woodland roadside near Slad, Gloucestershire. A tree has swung the Matador onto its side impaling Dick Barker's leg on a concrete post and it took a quarter of an hour to stop the engine. Later David's offer to winch the Matador upright with the estate Fordson Super Major Roadless and Auto-Mower winch was declined by the police who insisted on a proper Scammell recovery and crew who failed. Tongue in cheek a policeman invited David to try. He soon not only uprighted the Mat, but towed her off the road.

The following three photographs date back to approximately 1935 - representative of so many mills of the first half of the century. Men, horses and machines seething like ants about their daily tasks. Scenes of history and heritage, savour it while you may.

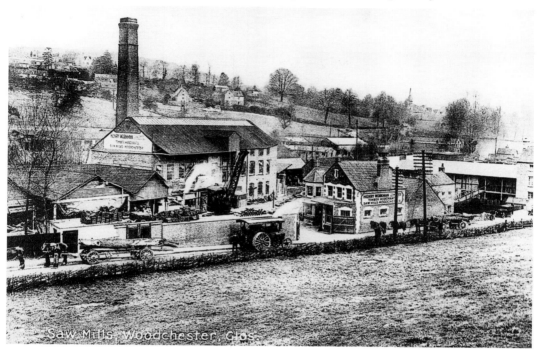

CHAPTER FOUR

Bill Friswell and Lionel Amos

Next I met Bill Friswell, student of the Lionel Amos school of 'Timber Haulingology' where the 'impossible' was the norm and miracles were more frequently required. Bill's education lasted about five years and stood him in good stead when he decided to run his own show, finding his niche in transporting mining timber, first with an Atkinson, of which he speaks well, then a Scania, for which he has even more praise. Both these vehicles came to know every road junction between Ludlow and London intimately. Just as it seemed Bill had 'cracked it' his hopes were dashed by the 1970 miners' strike. This was a fatal blow to this man's brave attempt to 'roll his own wheels'. Fortunately the ensuing heart attack was not. Nevertheless his HGV licence and plans to use his hard won skills vanished for all time – a cruel blow indeed.

One notable member of the 'stack 'em high and long brigade' was Lionel Amos, a name known across the county as well as that of Arthur Green of Silsden, Yorks (see page 35 of Stories of Round Timber Haulage) where Lionel learned his craft as a lad in the late 1940's. In the early 1950's he moved down to W M Chalk of Salisbury, hauling with their ERF from Ross-on-Wye to London. Later he joined Dennis Brown of Nailsworth. Young Lionel was a man of vision, more like double vision as he set himself almost impossible targets! In 1953 he bought a petrol engined Unipower and pole trailer, hauling from Lampeter in Wales to Chesterfield in Derbyshire. In the first six weeks he never used the choke once(!), since the engine was never allowed to get cold enough to need it and, understandably if you know Lionel, during that first six weeks this ambitious man had earned enough money to repay the purchase cost of this outfit. When the job ended he moved up to C G Perks of Ludlow. During the next ten years he built up a fleet of Artics, Allis Chalmers HD7 Crawlers, four ex-WD Coles Cranes, Matadors and FWDs to meet the demand of his expanding empire. In the mid 1960's he foresaw the timber trade's decline and commenced diversification into other businesses.

I first met Lionel in 1982 at his then home at Dursley, Gloucestershire. This was a fine, big house where William Tyndale is said to have translated the Bible into English, for which he was burnt at the stake in the 1500's. We were hurtling up the A38 in his white Mercedes with me interviewing him via the tape recorder in my lap. We sped up to a large bus garage at Epney by the River Severn to see the now famous ex-Jabez Barker JUX 135 1939 6-wheeled D G Foden Timber Tractor then entombed below a massive heap of bus seats, some of which Lionel moved to partially expose this sorry looking vehicle. Some would have concluded that the bus seats were worth more than the Foden, such was the sight that beheld me. Today it is more than probably the most photographed vehicle of the rally field, coupled to the last remaining Foden pole trailer. This triumphant restoration is now owned by his daughter, son-in-law and grandson. Other thoroughbreds off the Amos stable include a 1960 6-wheeled Foden S20 heavy haulage tractor 805 GRL 'Growler' and a 4-wheeled S21 'Mickey Mouse' Foden YUV 684. Although these vehicles are mainly for preservation some still work in the family woodlands as does their 1921 Dennings of Chard Rackbench saw. I last saw Lionel at the 1997 Merrist Wood Forestry show, brilliantly put on by the Round Timber Club. He'd driven 'Growler' from Epney and has been to Holland with these heavyweights, which at 67 years of age is not bad going. Arthur Green Ltd had a name for hard graft and big loads and it's been reflected in this man's own timbering. Lionel's philosophy was 'Keep the wagons well loaded". It was a matter of what the cart would carry, not what the horse could pull.

Lionel's first workshop and yard in Shropshire.
Note the roof mounted floodlights.

Not the world's strongest man contest!
Just a humorous illustration of the type required in the Amos team!

Lionel's ex Bulmers Cider ERF with Rolls Royce engine that drummed
in this type of cab according to Bill.

"The Growler" heavy haulage 1960 Foden tractor that Lionel drove to Holland.

Bill and his own Scania with timber from Perks of Ludlow proceeding to London.

This is the ex Jabez Barker famous Foden 6 x 4 timber tractor and Foden built pole trailer immaculately restored by Lionel Amos (insert top right) (Insert Bottom Left) Lionel moved a whole heap of bus seats in order I might photograph JUX135 as she awaited restoration in 1982.

The Atkinson well loaded with coverboard on her regular London run.

Bill has a fair load destined for NCB Chesterfield before the miners **"put the boot in"**.

Leyland "Hippo" driver Bert Lewis with perhaps 625 cube or so being towed out of a steep Welsh valley with an International TD9 operated by Bill Friswell bound for Perks of Ludlow in 1956.

Bill with ERF BUX 435 with heavy load of oak at Bromyard heading for Pontrilas sawmills in 1958.

CHAPTER FIVE

Bernard Owen & Jet Morgan

It all started when Ben Hinton rescued a set of Tongs, Dogs, or Timber Grapples as some call them; hefty hooks that when lowered onto a tree would grasp it for lifting. Laying in a Scrap Skip bearing the name of the maker, Bill Owen, a brilliant blacksmith employed by brothers Bernard and Jim Davies of Bucknell, Salop. Established kings of timber in those parts who came to employ 300 workers, some of whom were the toughest timbermen in their day, operating from a 5½ acre yard. This sort of heirloom was appreciated by his son Bernard and what you are about to read all stems from that gesture.

Some of my readers are already acquainted with this remarkable firm, B & J D (as I'll refer to them) through my previous books. On this special day I was to meet and mingle first hand with a breed of men all tarred with the same brush of 'dogged ability' and the Davies lads brought their share of glory and are part of its legend. Men who like others assembled, had daily smelled burning clutches, brakes and pounding hot engines, men and machines who had challenged the mountainous hillsides of Wales.

One was Bernard Owen who left school and served B & J D for over 23 years. He was to drive and operate the Davies futuristic trend setter, being the idea of mounting a James Jones hydraulic loader on an ex WD AEC Militant vehicle weighing 16 tons. A unique idea then but so common in our trade today. Bernard had a picture of a giant beech that yielded 1,000 cube of merchantable timber, even though it was so hollow a man could stoop and stand inside it. When he and his mate, Les Webster, were made redundant they became partners and now carry out tree planting, fencing, timber extraction with 2 Cat D4s, a County tractor and a Massey Loader. Les had a handful of Vicker Viscount scenes I featured in "Men Mud and Machines". I've seen a film of the 16-hour a day horrendous pounding test trials at Doughton Castle – visual proof of the nerve racking abilities and demands of this then powerful new crawler. Les was in the team, but otherwise mostly on extraction, often in terrain that called for every trick in the trade and often on hazardous slopes. Three other men who'd left and teamed up would also have had their fine stories to tell. They are all still together, but sadly in Bucknell cemetery.

One treat laid on for me was the arrival of ex Davies 6x4 Foden Reg GN622 timber tractor, beautifully restored to all her former glory and now owned by Les Morgan and previously driven by him and two of his brother during her murderous days of the 'Bucknell

Bash'! All contact was lost when the Foden was sold to a Scottish buyer. Then one day, almost unbelievably, Jon Garman was asked in a Ludlow pub if he knew of her current location and he did not. Then in a chance in a million, a complete stranger overheard the request and said he'd seen a B & J D Foden in a Scottish village scrap yard of dubious location. Jon drove up there and searched the area for a long time, but in vain. However, he was finally directed to the yard in question, bought and low-loaded it home after five years over the border. He stripped the Foden down so much so that when his living had to overtake his hobby time and he decided to sell the vehicle, the parts were collected on two small lorries by the buyer, John Onions of Shrewsbury, who carried out an excellent rebuild. A further mystery is that when completed and offered for sale, by slender chance Les Morgan heard of it, bought it and tells with great pride of the thrill and joy that he and many others have to see the Foden back home in Bucknell where she belongs. Les went to drive a JCB for someone for two weeks. Then, when a six week job lasted for two years, he bought his own JCB and dumper and 'never looked back' as they say.

The last doyen of the Davies Ultra timber time at this gathering of rugged Shropshire (at least one time) lads was Les's brother John, or Jet as they call him. Jet started on the above mentioned Foden, being one of three brothers to drive it.

Jet has the added gift of recalling umpteen asides that resulted in roars of laughter from the hospitality area. Like the time a driver cornered with a long load; putting his hand up to the shoppers he remarked on their solemnity to his mate, who replied "What do you expect, you've just shoved the front of the shop in!!" Once, when low loading the Vickers Crawler behind a 6-wheeled AEC Matador, they met rivals Jabez Barker hauling a big Wellington fir along a narrow road. Endeavouring to pass slowly, the Davies outfit rolled over into the ditch. Reporting the joint recovery led to red faces with both firm's employees. There was the farmer who claimed not to know how many sheep he owned yet would know if one was missing! Or the day Jet's pole broke at the 'swan neck', causing the whole back end to head for an oncoming postwoman. Taking quick, evasive action, he redirected it into a handsome set of wrought iron gates at a big house owned by a friend of the governor. Jet's explanation was not found acceptable, but requests for his dismissal were not taken up, although I think many an irreplaceable timberman has been sacked then instantly reinstated. One old ale swiller who lived in a row of identical houses had his doorstep whitened to ensure he entered the right house. He'd fall asleep, then awake demanding his supper. Sometimes as he slumbered, his wife would apply fat on his beard around his mouth, then say, "You've had it, look at the grease round your mouth" as the old fellow sat there in a drunken stupor. Of a garrulous lady, Jet said, "The only time her mouth wasn't open was when her eyes were closed". As I labelled hours of tape, I realised that day I had been dwarfed by giants who had a lust for life in timber. Truly tough conditions breed tough men. Men regarded as villains by environmentalists, and heroes by those who appreciate the risk to life and limb from everything nature throws at them: sub-zero temperatures, mud up to the crutch or blistering heat to harvest this renewable resource. Those that overlook the extent of creature comforts that come from the forest glad must remember that nature sets the price and these men and their ilk sometimes pay dearly for them. Getting such a diverse number of timber 'cutters and carters' (an expression still widely used) called for hours of planning. You reader, and I must be indebted to Bernard Owen, Jon Garman and Ben Hinton, all of whom made it possible.

The day they moved the village hall.

The custom built ex WD AEC Militant and James Jones mounted loader.
Operated by Bernard this whole unit was a new concept in loading and is still at work in Scotland.

B & J Davies had several Fowler "Challenger 3s".
Here Bernard is tushing out a big spruce in 1970; 600 cube in all - 127 feet long.

Big trees demand big tractors.

Hillside extraction road cutting was a frequent job for which Bernard and a Fowler were often needed.

This Davies brainwave was called "The Wizard". The back end of a Crossley was cut off and a Le Tourneau crane was grafted on. Re-engined with a Gardner 4LW driving the front wheels, the rear prop shaft powered the ten ton high lifting crane which lowered for transit.
Jet drove and enthused about it for some time.

A superb snapshot captures Jet in passing. Three Morgan brothers drove this Foden over the years. It was one of 9 purchased for various adaptations. All the jibs and anchors were made by Bernard's father.

Loading with "The Wizard".

Jet loading with a Davies built body AEC 6 x 6.

Grease up time for this Albion, hence the oil bucket and castor mounted creeper board.

The last Foden that Jet drove.

No, they've not stopped for Cherry Blossom. Here in Burton-on-Trent a big gun became unhooked from its army tractor, ran back down a hill smack into the Foden' offsidecab - the driver survived.

This AEC 6x6 even needs "Double Heading" approaching a steep hill with an enormous load.

The size of this enormous Onions Loggins Arch tested with a Vickers 180
by Davies at Dounton Castle is indicated by Les Webster and his mates below it.

CHAPTER SIX

Lawley Bros

Throughout the pages of my five previous books are glimpses of a whole host of men and women who had no alternative but to improvise to survive the rigours of daily battles with timber. Big bank managers often resulted in big improvisation. The brothers Joseph and Mark Lawley of Churchstoke, Powys, are no exception. Both ex Robert Wyne heavy haulage employees, Joe who at 75 is six days older than me (but appears six years younger) was hauling into Venables aged 18 and Mark is slightly younger. Joe's late wife, Joan, spent their first winter living in a van in Wales cooking, drying wet clothes, measuring timber, driving to Sandbach or Cannock for spares, a vast contrast to school teaching. As secretary, all the accounts, VAT, PAYE and wages were her job. Even when in the hospital maternity ward, Joe regularly turned up with a thick folder of work to be dealt with from the bedside. These two men teamed up in 1957 and seem gifted in 'vehicle grafting', Having a blacksmith for a father must have had some bearing I feel! How's this for starters?

When the crankshaft broke on their ex WD 15cwt Chev, the following transplant resulted. A Fordson Major, with gearbox, I emphasise, was grafted in with the engine protruding out front like a baby long-snouted Thornycroft! Unfortunately the gearbox drove the opposite way to the petrol engine, resulting in six reverse speeds and two forward, all complete with a Turner Winch. No, they didn't reverse it along the road, but often towed it. This was the first of many conversions to come and the start of two men set to succeed in round timber haulage. Mark recalled having over 200 cube behind a Standard Fordson. The 'stopping department' consisted of his mate riding on top of the load, leaping off and on to wind on the trailer brake, often at short notice. If you, like me, thought Lawleys had put an ERF cab on a Matador you'd be wrong. In fact Matador 4x4 running gear and transfer box were fitted and a remounting of the winch is what happened to this ERF, new in 1970 and converted in 1980. The jib is equipped with full swivelling rollers permitting roping at any angle, and avoids jumping off and down beside the top pulley, a luxury I never enjoyed. I talked with Brian, Mark's son, who drives the latest ERF. It seems two other Matadors had lost their former 7.7 AEC engines. One had gained a Gardner 6LW, the other a 150 of this marque, complete with David Brown gearbox. Yet another conversion is a Thornycroft Nubian Gardner 150 engined, of course. It had a most sturdy jib that had lifted a loaded parked-up grain trailer whose legs had gone through a culvert. Joe confessed he had used the timberman's stabilisers, a couple of blocks under the anchors!

Come 1968 the brothers were able to write cheques to purchase a new ERF at £3,000 and new BTC tandem axle pole trailer, price £1,000, such had been their capital efforts, total cost £4,000. One photo showed Brian with then shoulder length air at the wheel on a hefty load. The story goes that one of B & J Davies' drivers was criticised for a light load and chided with "You should see the loads Lawleys are carrying, and they've got women drivers". Another photograph I couldn't squeeze in was of a Land Rover being recovered from a 1:3 ravine. First the farmer's wife jumped clear, then he did. The two dogs who'd remained on board throughout were most reluctant to board another 'LR' that collected them all. In 1960 they had 300,000 cube, a year's work at Craven Arms. Every log had to be driven over a crossing on the main Manchester to Bristol line. Joe discussed the job with an inspector who okayed it and confirmed by letter, which he later wanted back since it seemed a higher authority had to approve. Canny Joe replied he'd returned the letter on receiving a further one, which never came. A CAT D6 with mud up to an ERF's headlights grossing at 32 tons was towing them out. The fan threw slurry up to behind the dashboard even, but so much mud and slurry was dropped on the crossing that when the trains roared through at 90 mph a railside cottage became plastered brown all over and Lawleys had to lay duckboards for the residents. At the end of the job the vehicle was sent back to ERFs for new brake linings and drums all round, price £475, but note, no other sign of wear and tear, i.e. stub axles etc. showing no sign of damage at all. Long loads became second nature to Joe who harped back to his previous Wynn days. When he had extra long ships' masts in Welshpool on a corner at one stage, he had just three inch clearance between the butts and a hotel, with his mate waving him on as the tails marked the walls of a bank. On the return journey a broken iron casting was observed, but he heard no more about it. Also when hauling 'longuns' from Ruabon to Wrenham, two policemen stopped him and queried the load length, of which he was unsure, so they measured it for him, claiming 35ft on the bolsters and 37ft overhang beyond. Uncertain of the legalities the 'boys in blue' asked if these lengths were normal and Joe replied, "Yes, all the time". This rather threw the constables who left the scene saying they would have to discuss it with their superintendent. Once again, benefit of the doubt was granted.

Another ERF tractor unit in the Lawley family is owned by Joe's daughter Megan. She uses it to transport his trailer mounted 1900 German built Kirchner Frame Saw to rally sites, plus a Ford Cargo she drove to Holland. Megan holds a Class 1HGV licence and CPC and recalled with zest how, when aged four, she would sit in Dad's lap helping to pull the crawler's levers and enjoying a sense of driving. Obviously I believe she really lived timber and was brought up in it. Leaving school, she dreamed of entering the family business, but father said "No, it's too dangerous!". All long before we paid lip service to H & S Regulations. On the rebound she joined the RAF for nine and a half years, where she became deeply involved in plane electronics and radar, known as avionics, I believe. Working on Buccaneers and Tornadoes across the UK and in Germany carrying Responsibilities with a capital 'R'. Her frustration is obvious since it seems her proposed career has to remain a hobby, but gender equality has moved on since that time and we must remember that Joe will have seen his share of timber accidents.

In my day woodland repairs were mostly carried out by 'crow bar and sledge hammer'. Today's fitters with all their snap-on tools have their limitations in this time of the electronic chip. One constantly hears of electronic engineers almost holding modern machine owners to ransom. Far be it from me to offer this young lady advice. But I'd say, go down and work in the woods by all means, not with a chain saw, winch skidder or 32 ton artic, all skills within her capabilities, but rather use those rare uncommon abilities gained in the RAF. Get your own 'Box of Magic' and set up a freelance mobile service! Forget the wheel in the woods, you've proved that on the road already. Go for it girl! Who knows, maybe your cousin would be glad to call you in.

Long loads became second nature to this family. The Lawley Brothers may have started out with more backward than forward gears, but diligence and acumen brought them success and finally a life more like that of fast forward in overdrive. This day I'd mingled with timber families who'd pioneered their art forms in the Shropshire 'neck of the woods' and I was indeed honoured to meet them.

The "Chev-cum-Major" transplant. Joseph and Mark prove necessity really is the mother of invention!!

Long trees being loaded by County Crawler on a P6 engined Austin artic.

Lawley's first ERF piled well and truly high in 1956.

ERF coupled to the new BTC pole trailer. A fantastic load indeed.

Shook, split and very heavy.

50 feet long oaks to cut into 12 x 12 valued at £1,000 a tree, bound for York Minster from Powys Castle. Brian is at the wheel as Mark reclines on the fuel tank.

A giant cedar at Powys Castle. Gosh what a girth!!

Joe and Mark sandwich an employee with the hair style of the day, as the head forester looks on and writes on the back of this marvellous picture "Big Oak at Llwynobin, Montgomery". Load on wagon 428 cube - 21 tons. Saleable timber of tree 827 cube, estimated weight of whole tree forty four and a half tons, approximate age 400-600 years. Circa 1960.

54

This snapshot doesn't reveal that these trees are *80 FEET* long. A special order for Cammell Laird of Birkenhead for which police co-operation would be required.

One of the re-engined Gardner Matadors. Look closely and note they're still topping up the ERF's hefty load.

Joseph Lawley aged 76 in 1998 brandishing his 60 year old axe backgrounded by the ERF 4 x 4. The chassis, engine and cab remains ERF mounted on Matador running gear. Lawleys replaced all Mat 7.7 engines with Gardners; even a 150 with a David Brown gearbox.

Megan Lawley with her own ERF (what else!). The make synonymous to Lawleys.

CHAPTER SEVEN

The Cundeys of Alfreton

A remarkable little engineering company that served the trade so well is that of Cundeys of Alfreton, Derbyshire. Founded in 1863 by Joseph Cundey, it continued through four generations until 1998 when partners Arthur and his sister retired. One contract their father, also named Joseph, and his Uncle William had was to supply and fit diesel and electric power units to imported German Bezner debarking machines. This led to their company developing the first all British machine in 1950 winning a silver medal forestry award in 1958. 1,800 were sold over the ensuing years.

Cundeys also bought ex WD vehicles from Ruddington. One batch was of Morris Commercial Amphibian 8 x 8 Terrapins powered and steered by variants of revs from 2 Ford V8 engines. The wheels were mostly adapted for Fordson tractors and Muirhill dumpers. A few subsequently went to river boards.

When Allinson Hodgson worked a traction engine nearby, Cundeys commenced repairs for him. The above trailer unit is believed to be a new custom built one off made for either of his two sister Fodens - two stroke 6 x 4s.

Allinson's fleet also included a Leyland Beaver Auto-Tractor and similar artic unit - a splendid line up.

One of their innovations was this Bedford QL mounted PTO driven hydraulic bolster loader which thrived until the advent of the current loader with abilities to load other vehicles.

An interesting photo of the Austin K6 (often P6 engined) pole wagon version of the hydraulic bolster loader, which had a K6 back bogie. The Austin K6 was very popular in timber.

CHAPTER EIGHT

Boys' Toys

Rugby Cement started quarrying chalk at Totternhoe, Bedfordshire in 1936 where plant included Ruston Bucyrus excavators, including a model 32-37-43 and two 54's, plus a USA Lima 802 with three Aveling Barford SL, 15ton dumpers equipped with crash gearboxes. Up to 70 were employed, despatching at peak five train loads of 30 27ton wagons, giving local Stanbridgeford railway station the highest turnover for its size in the UK; some 900,000 tons per annum.

For ten years I collected lengths of scrap cable for my various winches and carried out tree work occasionally through the quarry manager, Peter Trutch. He had taken a short-term apprenticeship with Ruston Bucyrus of Lincoln, where Peter was somewhat inhibited by having his father as Works Manager. Come World War Two he volunteered for that elite body of men, the Royal Engineer Muckshifting Company, where, true to form, they were banned from marching through a town due to their scruffy appearance! His 35 years with Rugby Cement commenced in 1948. This popular man excelled on employee safety and machine output, applying his own modifications with great success. Dumper brake drums that filled with slurry were targeted and an improvement to a famous cement plant maker's product won him this brief message from them, "We can't fault it or believe it".

At Kensworth, Peter devised alternator mountings for the two 54 R B's Paxton V6 engines to provide floodlighting for the night shifts. Another idea of his was dumper engine sound baffles. One fascinating innovation was the adaptation of an old Foden lorry for quarry service. First Peter extended the chassis, mounted a 500 gallon diesel tank, an air compressor with three reels of hose outlets for air, engine oil and fuel, a workbench, and even a walk-over gangway providing a most practical Workshop Wagon adaptation. When it came to any mechanical breakdowns of parts fitted, he got better results from manufacturers by referring to the parts concerned as suffering from 'premature failure', irrespective of being in or out of warranty, and resulting in discounts. At a loader demonstration, a 'whiz kid' operator shifted a massive heap of chalk from 'A' to 'B' like lightning, to be greeted with, "Now shift it back at a speed you could keep up all day"!!

As the quarry became 'worked out' in the early 1960s, a new site at Kensworth, six miles away, nestling below the Downs was opened so unobtrusively many locals are unaware of it. History tells us sheep roamed this area in the 1700's and do so again today on the green areas now reinstated, plus a marvellous Nature Reserve inspired by the Company. The gargantuan move was made over two years, since full production had to be maintained. Wynns transported the 43 and two 54 RB's with their front end equipment on low loaders, the gross weight of each machine being 78 tons. As for the Aveling loading shovel, and all the dumpers plus two Fordson tractors, Peter arranged for the local taxation office to issue a six mile exemption licence and drove the lot up there on temporary number plates which must have turned a few heads and saved money. Meanwhile, a colossal environmentally friendly operation was taking place at Kensworth. A ten inch pipeline had been laid 57 miles to Rugby for the ground chalk, converted to slurry by water, and pumped by three 750hp electric motors at the rate of 160 tons per hour, with NO booster pump en route. No pylons, trains or lorries! The two 250hp motors were on the Washmills. Remember, in the 1960's 'Environmental' was almost an unknown word. Peter Trutch, this old breed of 'hands-on' manager enjoyed a challenge and Kensworth quarry bears witness to it to this day.

I first met Harry Rutherford in the 1950's when I persuaded a farmer to hire me a Ferguson tractor with its front end loader tines replaced by a 'teaspoon size' bucket! The job of digging out a chalk bank and roots therein, was more suited to a Drott Shovel and brought about an uncontrollable roar of ear-splitting decibels! This did little to enhance the goodwill or the loaner. Harry's 36 years with Rugby Cement includes a wide range of skills from operating most excavators dumpers, of his day and being one-time quarry foreman. He had married into the Meakins family, all well grounded into quarrying and threshing. Aveling Barford dumpers from 15 to 35 tonners, powered by 400hp thirsty Rolls Royce and, later, Cummins engines were fitted with a one foot extension board around the skip top, which permits another two ton and are known as 'Greedy Boards'. One of the 54 RB's was assigned to Harry who remembers hairy moments like when he struck an unknown wet seam in the chalk face one night and tons of it spilled down above the tracks, forcing the jibing lever into action beyond control. Constant vigilance was required as big lumps of chalk could roll down the jib, through the window and into the cab – all perils of face shovel equipment.

As his 54 RB aged, the management took Harry to Ipswich to acquaint him with a new NCK Rapier excavator, similar in size to the 54 RB, but faster with air-assisted controls creating less fatigue and powered by a Cummins Trawler type of engine. The year was 1970 and the result was another purchased in 1978. With three foot wide tracks and an ability to dig eight or nine bucketfuls, i.e. 35tons in two minutes, the requirement of the quarry's demands for 6,000 tons a day was maintained. Up to that time drivers were doing their own maintenance and rope changing, but now this is let out to contract. The Kensworth family working atmosphere lives on in a team spirit prevalent today. My contact person for this story was Malcolm Tritton, a former youth club lad, now recently retired after almost his entire working life with the company who describes himself modestly as a quarry man. He's lived through the changes from 'Ropes to Rams' (hydraulic). From laying under dumpers before pits, driving them, and much more. His father, Tommy Tritton, had taken over one of the first 54 RB's at Totternhoe. Malcolm met me at the quarry where, adorned in a safety helmet and glasses, I boarded a Land Rover and was soon travelling the two miles or so road of dust clouds or wet slippery surface, dependant on the weather. Dwarfed like a Dinky Toy as a CAT 775D with her 60 ton plus load romped past keeping its 725hp down to the 20mph limit. The massive CAT 5080 face shovel was being serviced, but I did see the new CAT 375 Back Hoe bite into the chalk like butter and put 55 tons onto an Aveling dumper in a trice, as the CAT 988F loader with its nine yard bucket ambled by, all almost unbelievable to an old-timer like me.

My daughter Ruth had seen a lady from Kensworth Quarry present a cheque for £1,250 to the Dunstable and District Handicapped Persons Typing Club. This is a gesture made to various charities by the company, based on an accident-free four years at the Kensworth site. Quite wrongly I assumed this must be the quarry manager's secretary. It transpires she was, in fact, the **Manager!**

Emma Gough was born into a muckshifting family in 1972 in Devon. Even as a child she knew every machine's name and what it did, but was never allowed to drive them Leaving school, a neighbour took her and her mum round the Blue Circle Cement Works at Plymstock. Mum thought the place horrendous, but Emma saw this as her chance to become involved with 'Boys' Toys', as she calls them. This was to be her future calling; a dream that by sheer unswerving dedication is now a reality. At 18 she went to Hertford College at Oxford University to study Engineering Economics and Management, where she **was the first woman to gain a first class degree in this subject.** Leaving University in 1994 she joined Rugby Cement and has moved around (and up!) ten times, becoming Raw Materials Manager which includes Quarry Manager at Kensworth as well as the Midland Southam Quarry, which I imagine should include a degree in knowledge of the M1 as well! Now I have enough dregs of testosterone left to realise Emma is a most attractive young woman, yet as she related to Monkey Winches, 10 RB's and WW2 Cats, gender faded and she became a person with the spec, output and company production effects of modern plant I would hardly comprehend. Reeling off the fleet of Aveling Barford RD55's and CAT 775D Dump trucks, a CAT 508D face shovel, a CAT 140H Grader and CAT 375 excavator, all designed to produce approximately 1.9 million tonnes of chalk a year at Kensworth, and four CAT D300E artic trucks with two CAT 330 excavators, producing 0.6 million tonnes of clay/shale at Southam.

It was August 1999 and the whole of the processing plant was being renewed to pump 40,000 tonnes of chalk slurry per week against 23,000 tons per week as previous. The skyline of long crane jibs and construction of new plant movement had to be a nightmare itself. The whole system of management has changed today and Emma delegates to three Team Leaders: Martin Lithgo (Quarry Team), Vince McGuire (Maintenance Team) and Terry Davis (Production Team). However, ultimately the Manager 'calls the shots' and the 'buck stops' with Emma, who has established a sense of sex equality far removed from her first day when a nervous member of staff had enquired, "Will you want flowers on your desk?"

Chalk quarrying hereabout has put far more bread on the tables for far more years than most realise. Had a legacy of yawning stark gaps of mineral plunder, instead of restored greenery prevailed, hostile awareness would have abounded, I'm sure.

Most of the photographs in this chapter were taken by Malcolm Tritton, some especially for this book.

Harry Rutherford operating a 54RB in the 1960s.

Aveling Barford SN Dump Truck with crash gearbox - one of the first at Kensworth. Centre, well known all rounder Cliff Kent, right, John McCullogh.

Harry loading a SN No 14, 30 tonner fitted with a Allinson semi-automatic transmission. Peter Trutch designed the "baffle boxes" seen on this machine. The excavator is the much liked NCK.

Henry Sutton and Harry servicing one of the NCKs from the ERF mobile workshop that replaced the Foden when she succumbed to the rigours of the quarry.

Harry with his grandson dwarfed by a Cat 769C - one of 4 owned by the company.
He was driving this machine prior to this photograph being taken in 1997.

Aveling Barford (capable of carrying 55 tons) with John Pymm below and Gordon Dollimore
by the cab in 1997.

Cat 980B front loader with Sid, a popular fitter from Levertons.

Jayson Lapworth with the Akerman H16D Back Hoe - the quarry's first hydraulic excavator.

Aveling Barford RD 55 ton capacity. Note the two hot air ducts in the skip front that permit exhaust to heat the body to prevent chalk sticking.

The last 54 RB which had given 35 years faithful service in the hands of Tommy Tritton, Harry and others - scrapped in 1986. Bought, rebuilt and exported to India in "Pucker" condition!

The Cat 375 quickly loads 60 tons on to a 7750 dump truck - one of three now owned by the company.

This Cat 5080, new about 5 years ago, was one of the first of three built. One went to the USA, the second to Australia and this, the first in Europe, operated here by Rennie Miller.

A close up of the 5080 in action.

A forward view of the new 77 ton Cat 375 arriving in July 1998.

A close up of this super dirt digger!

Summer 1999. The new plant arrives. Leicester Heavy Haulage arrive with **PART** of the new wash drum - 70-80 tons of it.

A close-up of the massive, super crane capable of lifting 1,000 tons - one of only two in the country at that time, and believed to have hire costs on some sites of £7,000 a day! With it came a crew of 20, 3 lorries carrying the jib etc, another big mobile crane to assemble its big brother, and mobile kitchen for the crew.

Two lesser mobile cranes lift the new 75 ton crusher. I imagine Emma and the team leaders had some hectic daily planning meetings.

Emma Gough, Quarry Manager.

Quarry staff Barry Meakins left and Ken Tomlin right, support Emma as she presents the £1,250 cheque to Reg Hayward, Treasurer of the Dunstable Persons Handicapped Typing Club. June 1999.

CHAPTER NINE

From the Stone Age to Landing on the Moon

BSW Timber plc (Brownlie Smith and Western Softwood) formed in 1988 is a rather daunting title of a company for the likes of me to tackle, albeit briefly. Over 150 year ago a James Weddell had a small sawmill at Carluke, Lanarkshire in Scotland. James had a sister, Janet, who married local blacksmith John Brownlie. Two of their sons, Alex and Robert, inherited the mill and in 1856 moved to Earlston, Berwickshire – now the headquarters of BSW. Against fierce competition, Brownlie's gained a foothold in the burgeoning coal mining and railway markets.

In passing I mention one Alec Brownlie, a shrewd hard-nosed businessman who, when a forester questioned a failed job interview, bluntly replied, "Because you're useless, and I'm ruthless". Yet Alec had a unique common-touch between landowners who sold the trees and the men who felled them. Come WW2, in spite of price control and rationing, the company redoubled its efforts to meet Ministry timber demands and which in 1940 saw them tender £18,000 for one estate alone. Elsewhere the Timber Control Board required 650,000 cubic feet of standing timber to be felled by the Brownlie's. Before mechanical extraction, some 80 pairs of horses moved the trees and 300 men and women were employed. In 1949, at the Centenary Celebration, 600 people attended a dinner dance. Mrs Mary Brownlie, who'd been instrumental in keeping the firm going after her husband's early death in 1905, cut a cake, watched by her 13 year old grandson Sandy. Lack of space necessitates I jump to the fourth generation of this robust industrious family. Today the Company Chairman is John, younger brother of Sandy Brownlie CBE, the Chairman, who on reading my books, wrote me an appreciative letter accompanied with a video and book of the "Brownlie Story".

The recession of the 1980's decimated the UK timber trade and when our only three pulp mills closed down, mill after mill went to the wall. But Brownlie's survived, thanks to a diligent diversification. In Wales the twenty year old modern 'Western Softwoods' went up for sale. In the biggest decision ever made in the company's history, the Brownlie brothers took a highly calculated risk and bought this mill in 1981 and its super sister mill, a hive of modern technology, three days before the Prince of Wales opened it at Newbridge-on-Wye. A and R Brownlie were now set to be a leading force in UK softwoods. As many outlets for chips and waste disappeared, BSW found new ones abroad.

A substantial company, Thomas Smith and Sons, who also owned sawmills in West Scotland, had successfully approached Brownlie regarding a merger. The conversion of part of Europe's largest (130 acres) railway marshalling yard near Carlisle into the UK's largest sawmill at 30 acres. Banks of computer screens like a TV shop window scan the various operations. A far cry from Tom Smith's father's Rackbench saw, powered by a Chrysler car engine controlled by a piece of string pulled by the sawyer for added power on big logs! Today this mill devours 50 big loads of timber a day and despatches 25 loads of sawn timber. The BSW group embraces five of Britain's top sawmills, processing 800,000 tonnes of timber a year, now our biggest buyer of standing timber at 1.2 million tonnes annually. But these are brief background facts.

In 1914 an estimated 200 sawmills, mostly portable, set up in the woods of Scotland. Generally manned by 14 men with 10-12hp steam engines, producing 50,000 plus cubic feet a year. A rugged, dedicated, fabulous breed of men and women symbolised by the seven pound axe they wielded, sprang up. Some lived on site as families in stout, portable, sectioned wooden huts. Others would live in very basic caravans called 'Bothies', with just a four holed stove to meet all heating needs. Brownlie's had their share of these spartan dwellings and mills, and a name for long and loyal service with several generations of families following on. I also note a Trade Union once failed in its homework and demanded conditions less than were already being enjoyed.

They say, "It's an ill wind" and the two 'windblows' (violent storms) of 1953 and 1968 flattened entire forests, the latter alone taking four years to clear up. The bitter winter of 1946/47 stopped all operations for three weeks and 1962/63 for ten weeks. Let us, south of the border, imagine our worst conditions and realise this is generally the norm, north of it!! Stalwarts, like Jimmy Yule, was one of 30 in the Earlston Mill in 1938, earning twelve shillings and sixpence for a 50-hour week with only New Year's Day and a half day for Earlston Fair as holidays. According to one old hand, Bob Sawers, it took five years to be recognised as a Saw-miller. Another veteran, Charlie Elliott, coped with plant and machines from 1930 vintage to the later larger kit the brothers aspired to. A man named John Cannon was expert in tasks from tree valuation to complicated machinery installation. The longest serving man was David Sutherland, who clocked over 60 years, even returning from retirement in WW2 and was suitably awarded the British Empire medal. One old 'slogger', Doug Cameron, a sawyer, enjoyed what he called 'an aperitif', which consisted of 16 pints of beer on a Friday night.

The list of 'Brownlie Giants' is endless. Finally, two names that impress me are those of Jimmy and Isabella Fairbairn, a noteworthy husband and wife timber haulage team. Jimmy served Brownlie's for some 45 years and his wife about 20. She was first in the mill, cutting keys for railway lines and bobbins for export, plus helping on the firm's farm when required. Then when the Fodens came, she became legal obligatory loader and mate to Jimmy. Their son, John Fairbairn, an International haulage contractor, recalls how he lived in one of the sectional wooden homes for 14 years, moving from site to site and would drive a crawler through the school holidays. Isabella drove the Foden off the road, winching, loading, etc., whilst Jimmy did grafting that comes with 'Sheer Leg' loading. I've driven all the 4x4 timber tractors in dire conditions and applaud any woman who can handle a 2-wheel drive Foden where every foot of terrain requires foresight to traverse it. She could just about lift the trailer drawbar for coupling, but mostly propped it up at the desired height with a piece of wood.

These few great characters represent the valiant workforce in part of which BSW stands today. Sandy Brownlie had seen many small family firms merge and survive in Scandinavia and Canada, saw this as the way forward, and BSW was conceived. At the 1998 APF demo at Richmond in Yorkshire, I asked a young Scot if he'd any knowledge of the elusive Brownlie Fodens. It transpires he was in fact Sandy's son, young Alexander Brownlie, a member of the fifth generation, who'd just graduated in forestry from the University of Bangor. From the James Wedell acorn a mighty oak with diverse branches has grown. This chapter's title voices words spoken by Tom Smith Jnr describing the Carlisle mill and, as an old 'stoneager' myself, that sums up the parent company also - a century and a half of excellence and achievements.

This scene at Stobo station about 1920 neatly illustrates the importance to the timber business of both the horse and the railway as primary providers of transport. The horse-drawn pole cart with its solid tyres carries pine logs.

One of the portable sawmills set up midst portable homes where families of woodcutters and sawyers lived spartan lives miles from anywhere.

One of Brownlie's well loaded Leylands with solid rear tyres in the 1920s. On the back of this snapshot I note further copies are available from a chemist in Galashields price two and a half *OLD PENCE* each!!

Mining timber was a large outlet. Here we have horses, men and huge stacks of it drying.

Foreman David Dewar stands by an early Cat (perhaps an R4) in East Lothian. Note the sledge.

One of the firm's Leyland "Bull's" unloading at Earlston in 1946.

The transistion from the old ways to the new can be seen in this illustration, with an ex WD
Canadian Ford four-wheel drive vehicle, a horse and a man all combining to load Scots pine in the
Horse Shoe Plantations at Mellerstain in 1948.

Driver Jimmy Fairbairn with an Auto-Mower front mounted winch. Note Fordson engine with its long cranking handle. What a unique and splendid photograph. This first British company to make timber equipment was founded by George Grist of Norton St Philip near Bath.

Jimmy and his wife Isabella just complete loading a fine oak butt on rail with Foden HHT 366 at Gifford station in East Lothian. The practice of hauling big butts to a railhead for a week then returning home loaded was part of the job.

Jimmy preparing to load with shear legs.

Loaded with pride, this magnificent oak would be worth thousands of pounds today. Persons featured are Jimmy, his wife and woodcutter Maurice O'Brien.

Jimmy (who stands over six feet) and Isabella pause for a snapshot whilst tushing and loading big green beech trees in the 1950s.

Jimmy's son John spent hours helping Dad tushing out timber albeit aged 13-14 years old and he loved this old crawler.

Charles Elliot who, with his sons and joiner John Young, rebuilt this first registered 1942 Foden ***TWICE***!

Driver Jock Anderson's 1943 registered sister Foden HVT 854 sadly rots away in a scrapyard at Garmouth, Morayshire. A situation made worse because she was completely rebuilt by apprentices at the Foden works.

Extracting Douglas Fir in dire conditions with an International Hough Paylogger in the late 1970's.

A splendid load of Douglas Fir 12 x 12s - 36-40 feet long. Cut by J G D Munro for A & R Brownlie, bound for an Admiralty order. Incidentally, 1999 marked the bi-centenary of the birth of David Douglas born in Scone, Perthshire. He brought the first seed of this tree to Britain, a magnificent specimen of which stands at Dunkeld - at 212 feet high, our tallest tree!

Eck Redpath loads softwood on an ERF 1980. A & R Brownlie.

The first load of Baltic spruce logs imported from Estonia by BSW at Inverness docks in 1995.

The Caley's of Marham

My stories have ranged from large companies such as the Great BSW Conglomerate to the small firms, such as the tenacious if now tiny family firewood business in Norfolk. Three generations of Caley's of Marham, albeit now part time, for 50 years have fought the elements over a diverse area converting big trees others wouldn't consider into fuel for wood burning stoves on a one time round of 26 miles. Barry, his brother Peter, and his son Jodie (who at 24 sufferes from ME after glandular fever still tends the needs of this mixed stable of tractors today) operate a TEF 20 Ferguson, Massey 950, Marshall series 2, Fordson power major, a Nuffield, Bedford winch and a Matador. Grandfather Cyril at 86 still drives tractors, cuts logs with the Nuffield powered saw bench, chain-saws, and preaches on Sundays. I admire not only this man's physical and Christian strength but the whole family typifies our great timber pioneers of old.

The Caley line up - including particularly an ingeniously converted K Reg Bedford, MJ Model ex Vacuum Tanker (now re-cabbed with a DM by grandson Jodie, equipped with winch and jib. Inset: Jodie 24, Barry 52 and Cyril 86).

The End of the Road

It is said it's 'Better to Travel than to Arrive', very applicable concerning the ups and downs of life that reduce some to a heap of missing and clapped out parts. Yet I look back over the last 20 years of book research with great pride and joy having met and mingled with so many great men and women. There's the man who says my books are not thick enough!, or the woman who said "She'd never seen her husband read a book before". The eager man who could spell the word Scammell, but unlike me couldn't drive one. I'd misspelt this word. The avid reader on a life support machine for whom I made an audio loop of a few positive intense words, backed by favourite tractor sounds which he heard but couldn't respond to. Plus letters from Australia and across Europe, one of which opened 'My dear friend if I may call you so' plus a photo of my books on sale in Helsinki, Finland. I recall a sawmill where I saw a lad of low IQ learning to master a mobile crane despite jibes from normal staff that distracted him enough to demolish the saw doctors shop with a tree tail leading to instant dismissal. Returning he started the crane next morning as the irate manager reminded him he'd been sacked. In response the boy exposed belt wheals on his back saying my dad did this and says if you ever sack me again he'd do the same to you, hence instant re-instatement. One old driver said "Our boss never broke a promise, mind you that was only because he never made one"!

Three years spent with folk suffering from slight dimentia to these unaware who they were, let alone where they were, has been enlightening, being stroke-related memory illness mostly. I recall an ex-tank man void of all general conversation whom I milked the following experience from ! "I used to load a 38 ton 'Churchill' in pitch darkness with tracks over the sides of a Rogers trailer guided on by two mates walking backwards up the trailer each blowing their fags to make them glow if I got out of line. On another occasion, imagine two Diamond T's double heading you up a mountain pass, sitting on the turret praying the two drivers would change gear in unison or join the vehicles littered along the sheer drop below." Billions of memories are locked in damaged minds.

A gifted young mental Therapist named Jenny working under Doctor Evans the Psychiatrist that tends my mental situation each have made this book possible, "kick-starting" my ailing mind along the way. Some would say fifty years late for a "Nutter" that regularly winched trailer loads of timber up hills with an aged Latil's anchors jammed in the verge, even in Summer!

I passed my driving test on the 12-3-1940, 60 years ago, adjusting the brakes of my old Ausin 7 as the examiner appeared. I can't believe I once drove a load of timber through Birmingham having missed the ring road. Yet today I drive so slowly that I pass horses without decelerating and riders acknowledge what they think is my caution. Like some lorries perhaps a sign on the rear of my car inviting criticism of its driver could mischievously quote the phone number of the "poor" chaiman at Barclays Bank ensuring work for his diminishing staff. It's called Tempus Fugit!

Again as always I've tried to honour women and remind them that their liberation goes even further back than Emily Pankhurst, WWI and Germane Greer. Even in the Bible it is written, 'the king asked for those young people skilled in all wisdom, quick to learn understanding science and qualified to serve and that they be taught'. Daniel 1. 4 and 5. <u>Young People not just boys.</u>

So the time has come to say "Goodbye", the hardest word of all to utter. Thankyou for buying my books and videos. Together we've raised £17,000 for charity and I still look to meet my £20,000 target. Let me leave you with this thought, it's about "The Biggest Room in the World". No, not The Dome, it's "The Room for Improvement". The possibilities are enormous.

Our faithful old Unipower TPO899 in 1983 which we restored. 'Only the woodworm holding hands kept the cab together'! Short big butts were often carried on the anchors secured by the winch cable via a snatch block, wearing the anchor spades on the road somewhat!

When you come to the end of a perfect day (Sawmill version)

A sawmill man one morn got up
And found the sun was bright
His breakfast food, each plate and cup,
And everything was right.
He heard the morning whistle blow,
And heard the saws begin
Their singing in the vale below,
The day to usher in.

And then he wandered to the mill -
Found every man in his place,
And each one working with a will
And with a smiling face.
The logs came up without a hitch
To saws as sharp as swords;
Each cut produced a perfect flitch
Each flitch the best of boards.

And not a pulley slipped a belt,
And life was just a song;
The logs to lumber seemed to melt
and not a thing went wrong.
The morning mail some orders brought
And cancellations none;
In all the letters there was not
A kick from anyone.

All day the mill, from early dawn
Till night began to fall,
Kept working on and sawing on
Without a break at all.
At last the mill man homeward sped
Without a woe or care,
And, kneeling by his little bed,
He prayed this little prayer:

'O Lord, I know that some time I
Will have to perish too -
I know that some time I shall die,
for people often do.
Today we never spoiled a board
And everything went right.
If it is all the same, O Lord,
I'd like to die tonight.'

Anon (with kind permission of Sandy Brownlie)